ANIMAL SCIENCE, ISSUES AND PROFESSIONS

PHOSPHORUS FORMS IN ANIMAL MANURE AND THE IMPACT ON SOIL P STATUS

ANIMAL SCIENCE, ISSUES AND PROFESSIONS

Additional books in this series can be found on Nova's website under the Series tab.

Additional E-books in this series can be found on Nova's website under the E-book tab.

ANIMAL SCIENCE, ISSUES AND PROFESSIONS

PHOSPHORUS FORMS IN ANIMAL MANURE AND THE IMPACT ON SOIL P STATUS

ZHONGQI HE
AND
ZHENGXIA DOU

NOVINKA

New York

For permission to use material from this book please contact us:
Telephone 631-231-7269; Fax 631-231-8175
Web Site: http://www.novapublishers.com

NOTICE TO THE READER

LIBRARY OF CONGRESS CATALOGING-IN-PUBLICATION DATA

Phosphorus forms in animal manure and the impact on soil P status /
Zhongqi He, Zhengxia Dou.
p. cm.
Includes index.
ISBN 978-1-61668-980-3 (softcover)
1. Phosphatic fertilizers--Environmental aspects. 2. Farm
manure--Environmental aspects. 3. Soils--Phosphorus content. I. Dou,
Zhengxia. II. Title.
S647.H42 2009
631.8'61--dc22

2010023913

Published by Nova Science Publishers, Inc. † New York

CONTENTS

PREFACE

Increased knowledge of manure phosphorus (P) chemistry is needed to optimize nutrient recycling and to minimize adverse environmental effects associated with land application of manure. Part I of this volume reviews and discusses various manure P characterization methods including sequential fractionation, solution ^{31}P nuclear magnetic resonance (NMR) spectroscopy, enzymatic hydrolysis, and solid state spectroscopic techniques. The author then present a case study that compares P forms in conventional and organic dairy manure identified by solution and solid state ^{31}P NMR spectroscopy. The results of this case study suggest that the lability of P may differ between organic and conventional dairy manures; therefore specific management practices may be required for organic dairy manure. Part II of this volume focuses on the impacts of manure application on soil P status. For this part the author present and discuss the results of laboratory aerobic incubation and field studies to evaluate short- and long-term impacts, respectively. The lack of stable organic P in relevant extracts of the laboratory or field soils with manure amendment indicates possible degradation of stable manure P. Sequential fractionation of pasture soils reveals that the levels of labile P are more closely related to the number of years receiving poultry litter, whereas the levels of stable P are related to the cumulative amount of poultry litter applied. The two different patterns imply that the impact of manure application on the environment (including runoff potential) should be gradually factored in with increasing length of application.

INTRODUCTION

A significant amount of phosphorus (P) is present in animal manure. Land application of animal manure can benefit crop production by providing P and other nutrients. However, P runoff from agricultural soil due to over-application of animal manure may cause surface water degradation. Best management strategies are needed to optimize recycling manure P while minimizing adverse consequences of manure land application. The chemical composition of P in animal manure is one of the critical factors controlling P transport and bioavailability in soils. In the last decade, efforts have been made to increase basic knowledge of manure P chemistry by applying novel and/or modified advanced analytic methodologies. In this chapter we review and discuss the research on characterization of P forms in animal manure. We further demonstrate, using case studies, how this knowledge has improved our understanding of the fate and transformations of animal manure in agricultural soils.

Chapter I

CHARACTERIZATION OF PHOSPHORUS IN ANIMAL MANURE

I.1. SEQUENTIAL FRACTIONATION OF MANURE P

Sequential fractionation studies of manure P initially focused on P availability for plants. Using sheep manure, McAuliffe and Peech (1949) were the first to prescribe a fractionation procedure which involved extracting the sample with an alcohol/ether (3:1) mixture, followed by 5% trichloroacetic acid (TCA) extraction, and then ashing the residues. The corresponding P fractions were designated as lipid P, acid soluble inorganic and organic P, and residual P. This procedure was later used by Peperzak et al. (1959) and Barnett (1994a; 1994b) to analyze P in various manures such as dairy cows, feeder and finisher cattle, hogs, and broiler and layer poultry. Of total P in the manure samples, lipid P accounted for 0.4 to 2.1%, acid soluble inorganic P 35 to 63% and organic P 8 to 53%, with residual P ranging from 11 to 41%. The acid extracted P is considered available for plant uptake (Barnett, 1994a; 1994b).

More recently, a more sophisticated sequential fractionation procedure was adapted and utilized featuring increasingly stronger chemical solutions - deionized water, 0.5 M $NaHCO_3$, 0.1 M NaOH, and 1.0 M HCl, in that order - to quantify pools of P with varying degrees of solubility. This procedure was based on the Hedley (1982a) procedure for soil P fractionation but replaced the anion exchange resin extraction as the first step for soil P fractionation with deionized water for animal manure because the latter contains a considerable fraction of water soluble P (Dou et al., 2000; He and

Honeycutt, 2001). **Classes of P in manure are defined by the extractant's** removal from the manure matrix, i.e. H_2O-P, $NaHCO_3$-P, NaOH-P, HCl-P, as well as residual P (He et al., 2003a). The five P fractions are generally assigned to labile water soluble P, labile P adsorbed on crystalline surfaces, labile P associated with exteriors of Al and Fe oxides or carbonates, stable Ca-bound P, and recalcitrant P, respectively, but such assignments should not be considered as exclusive (He et al., 2003a). These functional P categories can be further divided into inorganic P (P_i) and organic P (P_o). The latter is generally calculated as the difference between P_i and total P (P_t) in an extractant. It has been argued (Haygarth and Sharpley, 2000) that some labile polyphosphate and P_o could be hydrolyzed during the assay and thus become part of P_i determined by the Mo blue method (Murphy and Riley, 1962). Thus, **some researchers prefer the term "molybdate-reactive P" or "soluble reactive P" (SRP) for the** P_i determined by this colorimetric method (Haygarth and Sharpley, 2000). Indeed, this term change only superfically improves the method. If the sponatanous hydrolysis is severe, a modified Mo blue method could be used to address the concern (Dick and Tabatabai, 1977; He and Honeycutt, 2005).

Although time-consuming, the sequential extraction procedure allows P to be categorically quantified, thereby providing a tool for investigating P transformations in animal manure affected by various environmental and/or management factors. Phosphorus forms in various animal manures have been investigated using this procedure, including from cattle (Ajiboye et al., 2004; Dou et al., 2002; He et al., 2004b; Sharpley and Moyer, 2000), swine (Ajiboye et al., 2004; He et al., 2004a; Wienhold and Miller, 2004); poultry manure/litter (Dou et al., 2002; He et al., 2004b; He et al., 2006c; Sharpley and Moyer, 2000); and other less common types of manure such as from deer, sheep, fox, and mink (McDowell and Stewart, 2005; Uusitalo et al., 2007).

The modified Hedley procedure for characterizing manure P follows the same principle generalized in Figure 1 but with variations in certain aspects. The ratio of manure and extractant (g:mL) differed depending on the investigators: 1:200 (Sharpley and Moyer, 2000), 1:100 (Dou et al, 2000; Wienhold and Miller, 2004; He et al, 2006c), 1:50 (He et al, 2004b; Uusitalo et al., 2007). Extraction time is 16 h in most studies. However, using a dairy and a poultry manure sample, Dou et al. (2000) found that P release into the water extract fraction was not greatly affected by shaking times from 1 to 16 h. Of the sum of water extractable P removed through repeated (multiple) 1-h extractions, 68% (dairy manure) and 75% (poultry manure) was released in the first 1-h extraction, which was designated as "readily soluble P" and believed to be most susceptible to potential environmental loss (Dou et al., 2000). As 1-h water-extractable P in manure has been proposed as an indicator of runoff potential of manure P (Kleinman et al., 2002), a 1-h extraction time in the sequential fractionation scheme could make the data from water extractable P in the sequential fractionation more compatible with those for runoff potential, particularly if the water:manure ratio and the condition of the manure (i.e, dried or fresh) are further unified. Vadas et al. (2006) have found that the water:manure ratio and whether or not manures are dried and ground or fresh make big differences in how much P can be extracted by water.

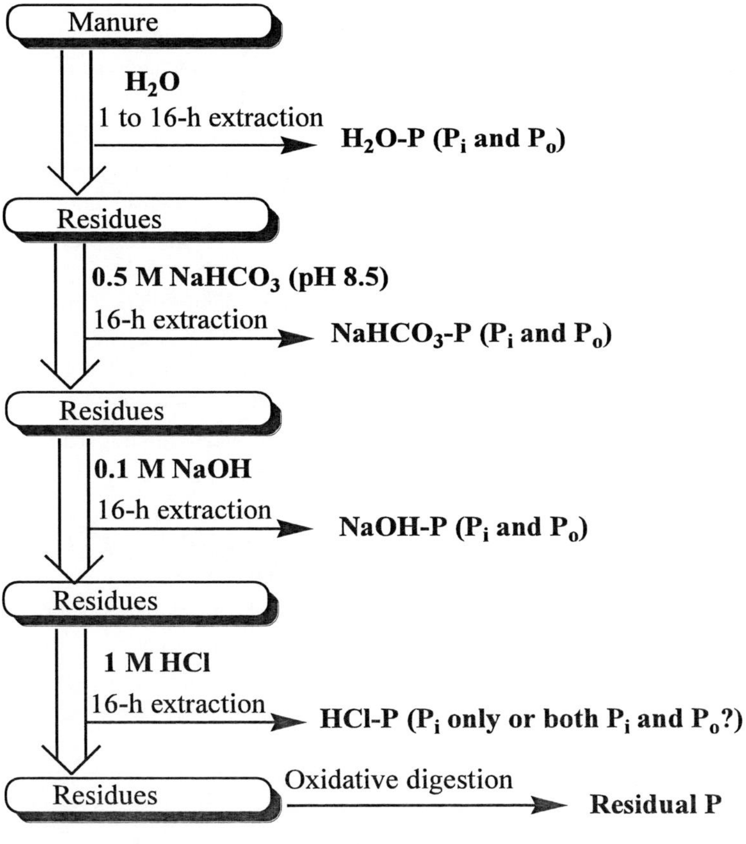

Figure 1. A typical modified Hedley sequential fractionation procedure for animal manure P extraction. Please see the main text for discussion.

The 16-h extraction time in other steps not only improves the extraction efficiency (Dou et al., 2000), it is also convenient operationally as a researcher can set up the extraction overnight. Unlike soil samples, animal manure extraction residues do not form tightly-compacted pellets during centrifugation. Hence, it is difficult to remove all clear supernatant (extracts) from the centrifuge tube. He et al. (2006c) suggested that a subsequent washing of the residues with 5 mL of water could reduce the **"contamination" of the supernatant leftover in the next extraction.**

There is concern regarding the forms of P (inorganic vs. organic) in the HCl extract. It is generally accepted that for soils, HCl extracts contain P_i only whereas P_o is negligible (Hedley et al., 1982b; Cross and Schlesinger, 1995). However, He et al. (2006f) demonstrated with 15 soil and 8 manure samples that some manure HCl extracts did contain significant amounts of organic P although other manure samples did not. Two manure samples even had P_o greater than P_i. Their data indicate that P_o in HCl fractions should not be ignored but determined experimentally. Without considering HCl P_o, any data using the sequential fractionation schemes might be incomplete.

Figure 2 displays P distribution patterns in four animal manures. If P in the H_2O and $NaHCO_3$ fractions is considered labile, more than 70% of total P in the dairy and swine manures was labile. In contrast, only about 30% of total P was labile in the poultry litter and fox manure. The labile P was primarily inorganic, whereas 54, 43 and 90% of P in the NaOH extracts were P_o in dairy manure, swine manure, and poultry litter, respectively. Organic P also accounted for 40 and 60% of P in the HCl fractions of swine manure and poultry litter. Phosphorus in fox manure had a pattern similar to poultry litter. Unlike poultry litter, there was little or no P_o in the NaOH and HCl fractions of fox manure. Foxes consume bone meal which has a high level of sparingly soluble P_i (Ylivainio et al., 2008). Even though P contents in animal manure vary greatly due to different management factors, such as feedstuffs, bedding, and storage type, similar distribution patterns have been observed for the same type of manure by different research groups. For example, the total P content of swine manure was 3.7 g kg^{-1} of dry matter in He et al. (2003a) and 19.8-33.6 kg^{-1} of dry matter in Wienhold and Miller (2004). The P in the swine manure in He et al (He et al., 2003a) was distributed as 47.2%, 19.1%, 18.3%, 12.2% and 3.1% of total P in the five fractions, very similar to the pattern of swine manure in Figure 2. Uusitalo et al. (2007) and Dail et al. (2007) reported similar distribution patterns for dairy and poultry manures, respectively. Therefore, best P management strategies should consider the specific manure P solubility (stability) in each type of manure.

The manure P fractionation based on a modified Hedley sequential procedure has been applied to study how dietary manipulation in animal feeding may affect P characteristics and their environmental implications. Using fecal samples from three independent dairy feeding trials and sequential fractionation, Dou et al. (2002) demonstrated that decreasing dietary P levels not only led to less excretion of total P in feces, but more importantly it reduced the amount and proportion of P that was water soluble. For instance, when dietary P was reduced from 6.7 to 5.1 or 3.4 g P kg^{-1} feed DM, water soluble P in feces decreased from 10.46 to 7.13 or 2.91 g kg^{-1} fecal DM, respectively. Water soluble P accounted for 83, 77, or 56% of fecal total P produced from the three diets. Other P fractions remained relatively small and were unaffected by dietary P concentration. These findings have significant environmental implications.

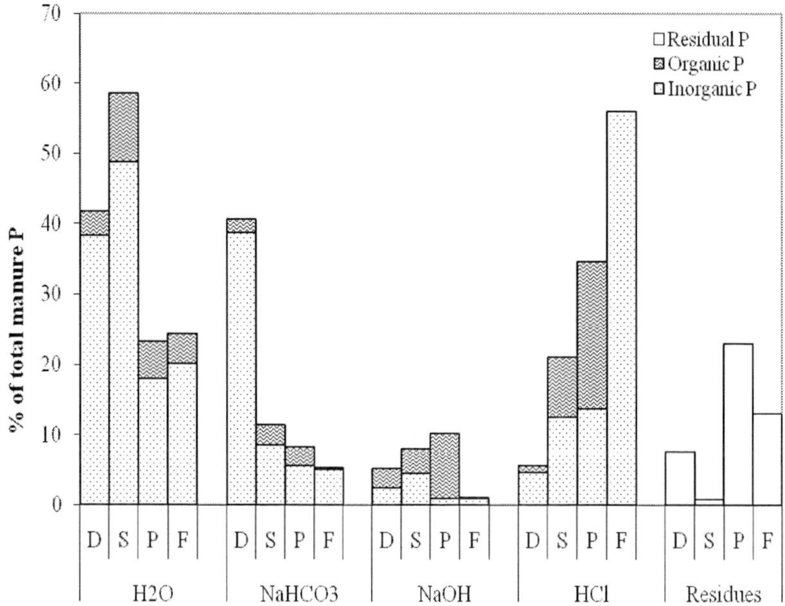

Figure 2. Phosphorus distribution patterns in sequentially extracted H_2O, 0.5 M $NaHCO_3$ (pH 8.5), 0.1 M NaOH, 1 M HCl and residual fractions of animal manure. Data are adapted from He et al. (2004b) for dairy manure (D) with 9.2 g P kg^{-1} of dry matter; Wienhold and Miller (2004) for swine manure (S) with 25.8 g P kg^{-1} of dry matter; He et al. (2006c) for poultry litter (P) with 22.3 g P kg^{-1} of dry matter; and Uusitalo et al. (2007) for fox manure (F) with 34.3 g P kg^{-1} of dry matter.

Water soluble P is most susceptible to potential losses through surface runoff and subsurface drainage. Reducing excess P in diets would markedly decrease potential environmental impacts. The P sequential fractionation procedure has also been used to investigate the efficacy of various amendments added to manure to stabilize P. For example, in a laboratory study, Dou et al. (2003) found that compared to unamended controls, aluminum sulfate [$Al_2(SO_4)_3$, alum] effectively reduced water soluble P in amended dairy, swine, or poultry litter manures. The reduction ranged from 80 to 99% at treatment rates of 100 to 250 g alum kg^{-1}manure DM. Fluidized bed coal combustion fly ash reduced water soluble P by 50 to 60% at a rate of 400 g kg^{-1} for all three manures, whereas coal flue gas desulfurization by-product reduced water soluble P by nearly 80% when added to swine manure and broiler litter at 150 and 250 g kg^{-1} but was ineffective in dairy manure. Another by-product, anthracite coal refuse fly ash, was ineffective for all three manures. Sequential extraction results indicated that the by-product treatments shifted manure P from water extractable P into a less vulnerable fraction, $NaHCO_3$-P, while the alum treatment shifted the P into even more stable forms, mostly NaOH-P. Such shifts in P fractions would have little influence on P availability for crops over the long term but may retard and reduce potential losses of P immediately following manure applications.

I.2. Phosphorus Forms Identified by Solution ^{31}P NMR Analysis

As ^{31}P is the naturally occurring phosphate isotope, solution ^{31}P NMR spectroscopy is an excellent tool for the study of P forms in animal manure. Leinweber et al. (1997) reported the solution ^{31}P NMR spectra of 0.1 M and 0.5 M NaOH extracts of swine and chick manures. More recently, NaOH-EDTA, adapted from a soil organic P extraction procedure (Bowman and Moir, 1993), has been used to extract P in animal manure for solution ^{31}P NMR analysis (Cade-Menun, 2005b). This method allows P to be grouped into six types of P forms: phosphonates, orthophosphate, orthophosphate monoesters, orthophosphate diesters, pyrophosphate, and polyphosphate (Table 1). Within these compound classes, several specific P compounds can be further identified, such as phytate (*myo*-inositol hexakisphosphate, IP6) in the orthophosphate monoester range, and DNA in the orthophosphate diester region. Although chemical shift values of many other P model compounds have been reported in the literature (Cade-Menun, 2005a; Makarov, 2005), caution is needed when attempting to apply those data to interpretation of the ^{31}P NMR spectra of "real world" manure samples, because ^{31}P NMR chemical shift values depend on the ionic strength of the solution, pH, and the presence of paramagnetic (such as Fe, Cu, and Mn) ions. In practice, identifying every P compound may be of limited use because the P classification presented in Table 1 should provide sufficient information with respect to the bioavailability or plant nutritional value of manure P.

Although solution ^{31}P NMR spectroscopy identifies the same P forms in different types of animal manure, the abundance of these forms is different among manures (Table 1). In NaOH-EDTA extracts of cattle manure, the majority of P is in the forms orthophosphate (62-70%), orthophosphate monoesters (15-28%), and phosphodiesters (4-11%). Pyrophosphate, polyphosphate, and phosphonates are also present with a total amount less than 10%.

Table 1. Major phosphorus forms in NaOH-EDTA extracts of cattle manure (CM), poultry litter/manure (PM), and swine manure (SM) identified by solution ^{31}P NMR spectroscopy

Form	Chemical shift (ppm)	Relative abundance (%)		
		CM[a]	PM[b]	SM[c]
Phosphonate	19 to 24	~ 1	ND[d]	NR[e]
Orthophosphate	5.7 to 6.1	62-70	34-66	86-94
Phosphomonoester:	3 to 6:	15-28:	30-66:	5-12:
Phytate		5-6	24-51	trace
Others		14-20	7-12	
Phosphodiester:	-1.0 to 2.5:	4-11:	2-7:	0-2:
DNA		1-5	0-1	
Lipids		2-6	2-6	
Others		< 1	<1	
Pyrophosphate	-4.5 to -5.0	1-7	0-1	0-2
Polyphosphate	-19 to -21	1-8	ND	<1

[a]: Data are from (He et al., 2007c; Turner, 2004); [b]: Data are from (He et al., 2007c; Hill and Cade-Menun, 2009; Turner, 2004); [c]: Data are from (Leytem et al., 2004; Turner, 2004); [d]: ND, not detected or less than 0.01% of extractable P; [e]: NR, not reported.

Table 2. Distribution of P in swine and cattle manures revealed by enzymatic treatment

P_o form	Swine manure fraction				Cattle manure fraction			
	H_2O	$NaHCO_3$	NaOH	HCl	H_2O	$NaHCO_3$	NaOH	HCl
Inorganic P	% [a]	%	%	%	%	%	%	%
Simple	61.1	38.9	18.5	88.3	57.8	52.5	35.0	92.4
monoester								
Phytate-like	2.8	2.0	35.0	1.7	0.3	0.5	9.8	1.3
DNA-like	15.2	2.0	4.2	2.5	7.2	0	0	0
Pyrophos-	2.1	0.7	1.5	- [b]	5.1	4.2	2.1	-
phate								
Nonhydro-	1.6	0	0.2	-	6.8	0	0	-
lysable	18.6	56.6	40.8	7.7	23.2	42.3	54.6	6.1

[a]: Percentage of P in each fraction; [b]: Not determined.
Adapted from (He and Honeycutt, 2001).

The NaOH-EDTA extracts of poultry manure are dominated by both orthophosphate (34-66%) and orthophosphate monoesters (30-66%) (He et al., 2007c; Hill and Cade-Menun, 2009; Turner, 2004). Phytate (24-51%) is the major orthophosphate monoester form (He et al., 2007c; Hill and Cade-Menun, 2009). Interestingly, recent reports (Leytem and Thacker, 2008; Leytem et al., 2004; Turner, 2004) showed that swine manure contains mainly P_i with trace phytate P irrelevant to the phytate content in feeds even though pigs are monogastric animals that are known for not being able to digest phytate P. Further research with various diets is needed to confirm the observation (Leytem and Thacker, 2008).

It should be kept in mind that NaOH-EDTA solution may be of low efficacy in extracting P from some manure samples, e.g. those with high Ca contents (He et al., 2007c). For example, NaOH (0.25 M)-EDTA (0.05 M) extracted only 48-70% of P in cattle, deer and sheep dung collected from an Australian grazing pasture (McDowell and Stewart, 2005). Turner (2004) reported a recovery of 80% of P from a cattle manure using the NaOH (0.25 M)-EDTA (0.05 M). With less than 100% recovery, caution must be taken when interpreting P forms in NaOH-EDTA extracts as whole manure P composition. He et al. (2007c) used an additional 1.0 M HCl extraction to recover the residual P after NaOH (0.25 M)-EDTA (0.05 M) extraction. Of the total P in their poultry manure samples, 65-71% was released in NaOH-

EDTA solution while 1.0 M HCl recovered an additional 33-34%. Solution ^{31}P NMR spectroscopy revealed that, of the HCl-extracted P, 22-23% was phytate, 4-5% was other monoester P and 0-1% was pyrophosphate P. This work indicates that the residual part of NaOH-EDTA extraction should also be characterized for a full understanding of manure P composition if NaOH-EDTA extraction does not recover all or nearly all P from a manure sample. Solution ^{31}P NMR spectroscopy coupled with sequential fractionation could be an alternative to provide comprehensive information on manure P composition and solubility (Ajiboye et al., 2007; He et al., 2008b; Turner and Leytem, 2004)

I.3. Characterization of Organic P by Enzymatic Hydrolysis

Enzymes are proteins that catalyze reactions to cleave specific chemical bonds. In nature, phosphatases catalyze the hydrolysis of certain organic P compounds to produce inorganic orthophosphate. Because organic P must be hydrolyzed to inorganic P prior to uptake by plants and microorganisms, enzymatic hydrolysis provides an estimate of the hydrolyzable, and thus the bioavailable, organic P in environmental samples, which is not possible to obtain with NMR analysis. He et al. (2009c) recently reviewed the principles and applications of enzymatic hydrolysis to investigate various environmental samples; therefore this section focuses only on the lability and forms of P in animal manure revealed by enzymatic hydrolysis.

He and Honeycutt (2001) first examined the release of P_o in sequentially extracted fractions of swine manure and cattle manure. About 49% of P in swine manure and 44% in cattle manure were present in organic forms, mainly distributed in the H_2O, $NaHCO_3$, and NaOH fractions. After pH adjustment, a portion of these fractions was incubated separately with alkaline phosphatase and wheat phytase, or incubated first with nuclease P1 or pyrophosphatase followed by wheat phytase. The enzymatic hydrolysis treatment did provide information about P forms found in the sequential fractions, in addition to the conventionally determined values of P_i and P_o (Table 2). The four designated types of P_o were distributed unequally in the sequential extracts. Phytate-like P was most prevalent in the H_2O fraction and accounted for 39% of P_o in swine manure and 17% of P_o in cattle manure. This finding is consistent with the lack of phytase enzymes in swine digestive systems (Wodzinski and Ullah, 1996). Simple orthophosphate monoesters comprised the major P_o form identified in the NaOH fractions. A small percentage of DNA-like P was found in each fraction (1.1-12%). Organic pyrophosphate was found only in the H_2O extracts.

As various approaches to phosphatase hydrolysis have been used to evaluate the lability of organic P in environmental samples, the diversity of approaches has complicated data comparison (He et al., 2009c). A unified approach for enzyme hydrolysis would allow data comparison across a range of P sources and forms. He et al. (2004a;2004b) have developed a stepwise approach for this purpose. In this enzymatic approach, organic P extracted from manure or soil is incubated at pH 5.0 (100 mM sodium acetate buffer) with (i) acid phosphatase from potato; (ii) acid phosphatase from potato and nuclease P1; or (iii) phosphatases from potato and wheat germ plus nuclease P1. Each of the enzymes shows optimal activity at 37°C and near pH 5. The P differentially released by these three enzymes is designated as simple monoester P, polynucleotide-like P, and phytate-like P, respectively. Another advantage of the modified method is that possible errors due to incomplete substrate overlaps by separate incubations with an individual enzyme are minimized by the stepwise addition of enzymes.

This enzymatic approach, coupled with sequential fractionation, provides a novel tool for understanding manure P chemistry and its impacts on soil P levels (He et al., 2004c; He et al., 2006c; Waldrip-Dail et al., 2009). A significant finding for application of this coupled method is that HCl fractions in the Hedley fractionation may contain both P_i and P_o , rather than P_i only as assumed in some of the literature (He et al., 2006f). In the original procedure proposed by Hedley et al. (1982a; 1982b), no P_o was included in the HCl fractions because the amount of P_o in the HCl fractions of their samples was negligible based on preliminary measurement. Since then, many researchers have simply followed the procedure and assumed that no P_o was present in HCl fractions without any measurement (Cross and Schlesinger, 1995). In practice, the inconsistent and/or different background interferences between P_i and P_t measurements made estimation of concentrations of P_o less accurate, even sometimes negative, as P_i is generally measured by a molybdenum blue colorimetric method, and P_t is measured using the same method after oxidative digestion, or by an inductively-coupled plasma method with or without digestion (Ajiboye et al., 2007; Codling, 2006; He et al., 2006f). Addition of enzyme(s) basically keeps the medium matrices of extracts unchanged. Thus, increases in P_i determination after enzyme incubation should be attributed to the hydrolysis of P_o by the enzymes added to these extracts. Using enzymatic hydrolysis, He et al. (2006f) measured P_o in 15 soil and eight manure samples. Their data indicated that, whereas the HCl fractions of some samples contained negligible amounts of P_o, others contained significant amounts of P_o. The concentrations of P_o were greater than P_i in one soil sample and two manure samples. Apparently, it should be determined experimentally whether the HCl fraction of a sample contains P_i only or both P_i and P_o. Without considering the HCl P_o species, any data using the Hedley sequential fractionation schemes might be incomplete or less accurate. Inclusion of HCl-extractable P_o with other sequentially extracted P forms would provide more accurate and comprehensive information on the fate and interchanges of P forms in soil and animal manure under various environmental conditions (He et al., 2006f).

Since both ^{31}P NMR spectroscopy and phosphatase hydrolysis are valuable tools for P characterization, He et al. (2007c) quantitatively compared the P forms in NaOH (0.25 M)-EDTA (0.05M) extracts of fresh and dried dairy and poultry manure determined by the two methods (Figure 3). Enzymatic hydrolysis identified polynucleotide-like phosphodiesters with abundances of 6.4, 3.6, 3.4 and 4.3% of total extracted P in the four manure samples, respectively. The abundance quantified by solution ^{31}P NMR spectroscopy was 7.7, 7.2, 4.2, and 3.4% for total orthophosphate diester P, and 1.3, 0.8, 0.9, and 1.6% of total extracted P in the four manure samples for DNA, respectively. These data are reasonable since the abundance of polynucleotide-like P should be greater than that of DNA but less than that of total orthophosphate diester P. The other three types of P identified by the two methods were more similar (Figure 3). Although the P forms and quantities detected by the two methods were not always the same, comparison of the two sets of data indicated that the patterns of P distribution in these animal manure samples determined by the two methods were similar, with a correlation coefficient of 0.937 (Figure 3). He et al. (2008b) further examined the differences in solution ^{31}P spectra of sequential fractions of poultry litter before and after enzymatic hydrolysis. The peaks of P_o species became very weak or disappeared in the solution ^{31}P NMR spectra of the enzymatically-treated fractions, which confirmed the validity of P identities in the untreated fractions detected by enzymatic hydrolysis or solution ^{31}P NMR spectroscopy. These observations suggest that both methods can be used to characterize P in animal manure and it is acceptable to compare the information on P data obtained by the two methods although some discretion should be applied. Researchers can utilize the method that best fits their specific research goals, or use both methods to obtain a full spectrum of manure P characterization.

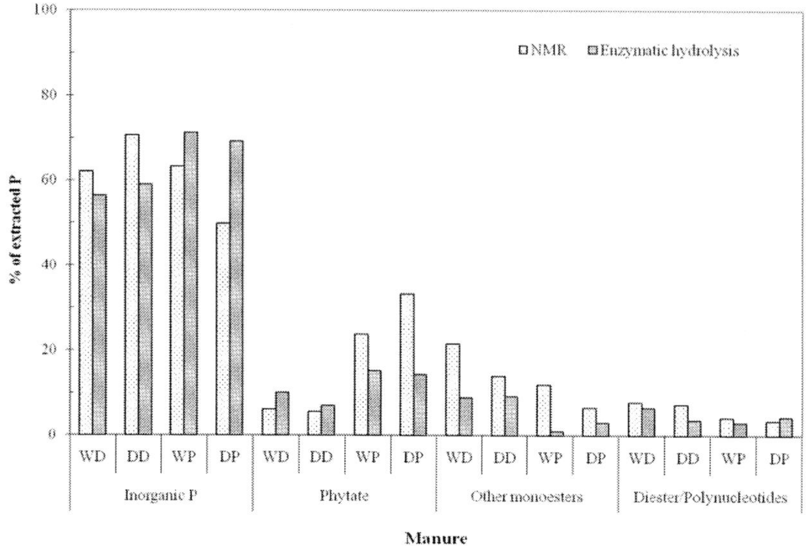

Figure 3. Comparison of phosphorus forms in NaOH (0.25 M)-EDTA (0.05 M) extracts of animal manure identified by enzymatic hydrolysis and solution ^{31}P NMR analysis (He et al., 2007c). WD-wet dairy manure, DD-dry dairy manure, WP-wet poultry manure, DP, dry poultry manure. The correlation coefficient of the two sets of data is 0.937 with the statistic significance at α <0.001.

The NaOH extractant used in either the sequential fractionation or solution ^{31}P NMR spectroscopy could spontaneously hydrolyze some labile P_o compounds (He et al., 2003b; Pant et al., 1994). To obtain more accurate information on labile P_o, He et al. (2003b; 2006d) separated a manure sample into soluble and residual parts under mild assay conditions using water and sodium acetate buffer (pH 5.0). The labile P_o in the two parts was then released by enzymatic hydrolysis. Using the mild assay method, He et al. (2003b) investigated the effects of storage on inorganic P (P_i), enzymatically hydrolysable organic P (P_e), and non-hydrolysable organic P (P_{ue}) in cattle manure and swine manure. Their data indicated that the major change during the storage of animal manure for a year was the increase in P_{ue}; thus, manure P solubility may increase with storage, but the increase would not produce more bioavailable P in the manure. The mild assay conditions could be further modified to investigate specific P compounds. For example, the reducing agent sodium dithionite has been included in the incubation mixtures to explore reducible Fe(III)-associated P_i and P_o in poultry litter and dairy manure (He et al., 2006a). Dao et al. (2006) directly added EDTA and fungal phytase to diluted dairy manure to characterize mildly (mainly Ca and Mg) complexed P_i and P_o. The simplified method can also be used as a reliable and convenient analytic method to directly quantify the phytate concentration in animal manure (Dao and Hoang, 2008; He et al., 2009a), since the use in feeds of low-phytate grains and supplementation with phytase enzymes may result in more unpredictable phytate concentration in animal manure (Leytem and Thacker, 2008; Maguire et al., 2004; Wienhold and Miller, 2004).

I.4. Metal Speciation of P Identified by Solid State Spectroscopic Techniques

The forms and lability of P in animal manure have been investigated by solution-based techniques as discussed above, in which the metal speciation of P was estimated by the properties of the solution and/or the extractant. Recently, solid-state-based advanced spectroscopic techniques, mainly solid state ^{31}P NMR and X-ray absorption near edge structure (XANES) spectroscopy, have been used to characterize the metal species of P in environmental samples (Ajiboye et al., 2007; Gungor et al., 2007; Hunger et al., 2005; Jayasundera et al., 2005). Those efforts have mainly focused on metal species of inorganic phosphate. Hunger et al. (2004) characterized the solid state ^{31}P NMR spectral features of poultry litter amended with alum, finding a mixture of poorly ordered wavellite and phosphate surface complexes on aluminum hydroxide that had been formed by alum hydrolysis. Furthermore, phosphate associated with Al comprised 40% of P with 14% of P as calcium phosphate phases in the alum-amended poultry litter. Shand et al. (2005) reported the main peaks and spinning sidebands of dicalcium phosphate dihydrate and ammonium magnesium phosphate hexahydrate in the solid state ^{31}P NMR spectra of sheep feces. Jayasundera et al. (Jayasundera et al., 2005) proposed the presence of various P_i and P_o compounds (including P_i complexed with Ca) in dairy manure samples they investigated by solid state ^{31}P NMR spectroscopy.

There are more published manure P research studies using XANES spectroscopy than solid state ^{31}P NMR spectroscopy. Shober et al. (2006) applied XANES fitting processes to identify various P species present in biosolids and manures, such as hydroxylapatite, phosphate sorbed to Al hydroxides and on ferrihydrite, β-tricalcium phosphate and phytate. Phosphorus K-edge XANES spectra of poultry manure showed no evidence of crystalline P minerals but instead showed the dominance of soluble Ca-P species and free and weakly bound phosphates (Sato et al., 2005). However, diets incorporating modified P minerals could change P speciation in manure: less soluble hydroxylapatite was observed and the aqueous phosphate contribution disappeared in P K-edge XANES spectra of turkey manure with Ca to P ratios greater than 1.7 produced from modified diets (Toor et al., 2005). Using X-ray powder diffraction and XANES, Gungor et al. (2007) determined P speciation of raw and anaerobically digested dairy manure with an emphasis on the Ca and Mg phosphate phases. Linear combination fitting of the XANES spectra showed that 57 and 43% of P in the raw manure was associated with anhydrous dicalcium phosphate and struvite ($MgNH_4PO_4 \cdot 6H_2O$), respectively. In the anaerobically digested manure, 78% of P was present as struvite and 22% of P was associated with hydroxylapatite, suggesting P speciation shifts toward Mg and the least soluble Ca orthophosphates following anaerobic digestion.

Whereas these two solid state spectroscopy methods have provided novel insight into metal P_i speciation, in studies where researchers have used one or two phytate compounds (i. e. sodium phytate $Na_{12}IP6$ and/or Ca hydrogen phytate $CaH_{10}IP6$) as reference compounds to study phytate, these two techniques were not able to distinguish phytate compounds in environmental samples (Beauchemin et al., 2003; Peak et al., 2002). He et al. (2006e; 2007a; 2007b) prepared different metal phytate compounds, and studied their solid state ^{31}P NMR and P K-edge XANES spectral features. These spectra demonstrated significant differences in the shapes and positions of spectral peaks, indicating these two spectroscopic techniques could be used to advance our knowledge of metal speciation of phytate in animal manure. In addition, Kruse et al. (2009) recently reported the P $L_{2,3}$-edge XANES spectra of a wide range of reference P compounds. Their results demonstrated that these P $L_{2,3}$-edge spectral features are richer than those of the P K-edge spectra. Therefore, the application of P $L_{2,3}$-edge XANES spectroscopy may yield more reliable results on P speciation in animal manure.

I.5. Case Study: P Forms in Conventional and Organic Dairy Manure

Organic dairy production has increased rapidly in recent years. Organic dairy farms have significant differences from conventional counterparts, including fewer imports of protein and energy feeds, a higher proportion of forage crops in the ration, and an increased reliance on manure and compost as nutrient sources, rather than chemical fertilizers. Although these differences could impact the turnover and availability of manure P to plants, little information is available to aid organic dairy farmers in making nutrient and manure management decisions. For this reason, He et al. (2009d) comparatively characterized P in organic (OD) and conventional (CD) dairy manure by multiple extractions, solution and solid state ^{31}P NMR spectroscopic techniques. Phosphorus in both types of dairy manure was separately extracted with water, Na acetate buffer (100 mM, pH 5.0) plus 20 mg Na dithionite mL^{-1} (NaAc-SD), or 0.025 M NaOH with 50 mM EDTA (NaOH-EDTA). The first extractant is for the most labile P (Kleinman et al., 2007; 2002), the second has a greater affinity for Fe-associated P (He et al., 2006a), and the last one extracts stable metal-P species (Cade-Menun, 2005b; He et al., 2007c). Whereas the three extractions gave data on solubility (i. e. lability) of manure P, further characterization of the extracts and residues by solution and solid state ^{31}P NMR spectroscopy provided information about the P species in the manures.

Table 3. P in extracts of conventional (CD) and organic (OD) dairy manure by H_2O, 100 mM sodium acetate buffer (pH 5.0) with 20mg/ml sodium dithionite (NaAc-SD) and 0.05M NaOH-0.025M EDTA (NaOH-EDTA) by solution ^{31}P NMR spectroscopy

Sample	P_t	P_i	Phytate	Other Mono-esters	DNA	Other Diesters	PyroP	PolyP
	---------------------------- % of total manure P -----------------------------							
CD- H_2O	50.3	41.1	0.0	5.9	0.8	2.5	0.0	0.0
CD-NaAc	88.2	61.9	6.4	9.3	1.4	1.4	5.6	2.1
CD-NaOH	81.9	55.9	7.5	11.1	1.2	2.5	1.2	2.5
OD-H_2O	42.4	37.5	0.0	2.8	0.3	1.7	0.0	0.0
OD-NaAc	94.9	73.2	10.0	6.6	0.9	1.7	1.7	0.9
OD-NaOH	96.9	77.2	9.5	6.3	0.8	1.6	1.6	0.0

adapted from (He et al., 2009d).

Whereas water extracted more P from CD (50.3%) than OD (42.4%), NaAc-SD and NaOH-EDTA were more efficient in extracting P from OD than CD (Table 3). These data imply that there is both readily labile and recalcitrant P present in CD, but more moderately labile P in OD. Solution ^{31}P NMR spectra show seven types of P species present in both types of dairy manure (Table 3). Orthophosphate (i. e. P_i) accounts for 70 to 89% of extracted P in these extracts. Polyphosphate and pyrophosphate are not present in water extracts, but in both NaAc-SD extracts, and in the NaOH-EDTA extracts of OD. As NaAc-SD extracted significantly more Fe than did NaOH-EDTA; this suggested that polyphosphate and pyrophosphate are associated with Fe in these samples (He et al., 2009d). Recently, Dou et al. (unpublished data) also observed greater percentages of pyrophosphate and Fe in NaAc-SD than in NaOH-EDTA extracts of two dairy manure samples. These observations demonstrate that the mild NaAc-SD solution is an alternative to NaOH-EDTA for extracting P from samples where P is more closely associated with Fe and hydrolysis of condensed P compounds is a concern.

In the solid state ^{31}P NMR spectra of the unextracted manures, a sharp peak appeared at 6.2 ppm with a broad peak centered at 2.6 ppm with minor spinning sidebands (He et al., 2009d). In contrast, the spectra of the residues of water and NaAc-SD extractions showed a major asymmetric complex peak at a region from -0.2 to 1.7 ppm with stronger spinning sidebands. The strength of spinning sidebands is a critical factor distinguishing organic and inorganic compounds with similar major isotropic chemical shifts: strong spinning side bands are observed only in the spectra of metal phytate compounds, minor spinning sidebands in the spectra of hydrogen inorganic phosphate compounds and no detectable spinning sidebands in the spectra of tribasic phosphate compounds (He et al., 2007a; 2007b). With these spinning sideband features and deconvolution of the major isotropic chemical shift peaks, various Na/K, Ca, Mg, Zn and Al species of P_i and P_o compounds have been identified from these solid state ^{31}P NMR spectra. Combining the data from the multiple extractions and solution ^{31}P NMR spectroscopy, 15 P species have been identified in these two manure samples (Table 4). Thus, He et al. (2009d) concluded that their conventional dairy manure sample contained relatively higher contents of soluble P_i species and recalcitrant metal P species. In contrast, their organic dairy manure sample contained more moderate soluble (labile) Ca- and Mg-P species. As these moderate forms would gradually release the nutrients, organic dairy manure could act as a slow-release fertilizer, thus increasing the efficiency of organic dairy manure as a nutrient source while reducing the potential of runoff into nearby surface or groundwater sources. However, further testing is required, since only one organic dairy manure sample has been studied to date.

Chapter II

Impacts of Manure Application on Soil P Status

II.1. Case Study: Laboratory Aerobic Incubation to Evaluate the Short-Term Impacts

Laboratory aerobic incubation has been used to evaluate the mineralization and bioavailability of N and P in soils amended with animal manure (Abbasi et al., 2007; Eghball et al., 2005; Griffin et al., 2005; Preusch et al., 2002). Sequential fractionation can be used to monitor changes in soil P during the incubation experiments (Buehler et al., 2002; Qian and Schoenau, 2000) and more information on labile P_o can be obtained when enzymatic hydrolysis is coupled with the sequential fractionation (He et al., 2004a). With this coupled approach, He's group conducted laboratory aerobic incubation experiments to evaluate the short-term impacts of dairy and poultry manure addition on soil P status (He et al., 2004c; 2006b; Waldrip-Dail et al., 2009). They measured P_i, P_{ho}, and P_{nho} (enzymatically hydrolyzable and nonhydrolyzable P_o, respectively) in the sequentially-extracted fractions of manure-amended soil at different incubation times and analyzed their patterns of change over time. Since the mechanism of interconversion of P species can be complicated, He et al. (2004c) assumed that interconversion of P species among the different fractions is mainly a physicochemical process and transformation of different P species (P_i, P_{ho}, and P_{nho}) within in a given fraction are mainly mediated by biological and biochemical processes. On the basis of this assumption, He et al. (He et al.,

2004c; 2006b; Waldrip-Dail et al., 2009) assessed the P dynamics of two Maine soils in response to dairy and poultry manure application.

An unnamed sandy loam and a Lamoine silt loam were used in the laboratory aerobic incubation experiments, with 11 dairy manure samples plus two chemical fertilizer controls (NH_4Cl and KH_2PO_4) (He et al., 2004c; 2006b). The P species in the H_2O, 0.5 M $NaHCO_3$ (pH 8.5) and 0.1 M NaOH extracts of soils were quantified over the 108-day incubation period. P in 1.0 M HCl was not determined, since these dairy manures contained only a very small amount of HCl-extracted P (He and Honeycutt, 2001; He et al., 2004a); therefore,this fraction should not significantly impact soil P dynamics. Data from this incubation experiment demonstrated complementary fluctuations between P_i in the $NaHCO_3$ and NaOH fractions as well as between labile P (P_i and P_{eo}) and P_{nho} within the same NaOH fraction. He et al. (2004a) proposed that these complementary fluctuations could be an important mechanism for maintaining the balance between labile and immobile P in soils. Furthermore, the general distribution patterns in manure-amended soils for the P_i, P_{ho}, and P_{nho} fractions followed the trend of changes for the inorganic fertilizer-amended (control) soils, indicating that soil properties played a major role in controlling P dynamics. In the H_2O and $NaHCO_3$ fractions, soil bioavailable P (P_i and P_{eo}) was linearly related to the added P. However, NaOH extractable soil P was not closely related to the P applied in the manure. In other words, a single application of dairy manure impacted soil bioavailable P in a manner similar to fertilizer and did not significantly impact soil P properties and the environment. This means growers converting from conventional to organic practices (i.e. where manure is the sole source of N and P) may estimate manure bioavailable P to meet their short-term P needs whereas the impact of manure application on soil properties and environment should be gradually factored in with increasing years of application (He et al., 2006b).

A significant portion of P in poultry manure/litter is HCl-extracted P, which differs from the relative P fraction ratios of dairy manure (Figure 2). Thus, the aerobic incubation of poultry manure-amended soils provides information on the fate of the stable HCl-extracted P in poultry manure upon application to soils (Waldrip-Dail et al., 2009). Overall, poultry manure amendment affects the soil P status primarily in the H_2O- and HCl-P_i fractions. As observed with dairy manure, soil properties, including soil mineral phase and sorption-desorption processes, are key factors in controlling P dynamics in soils amended with poultry manure. With up to 58% of extractable P in the HCl fraction of the poultry manure, which had relatively equal amounts of P_i and P_o (Dail et al., 2007), the P_i level in the

HCl fraction of poultry manure-amended soils remained relatively stable over the course of the incubation. In contrast, the P_{nho} fraction rapidly became enzymatically hydrolysable, with a portion remaining in the HCl fractions and some transformed to other fractions. This observation indicates that the stable HCl-extractable P in poultry manure has rapidly degraded to more labile P forms in these amended soils. Whereas sequentially extracted HCl P in soil is considered less plant available (Cross and Schlesinger, 1995), the incubation data from this experiment confirmed that this may not be true for the HCl-extracted P in animal manure (Dail et al., 2007).

Table 4. P species in the conventional (CD) and organic (OD) dairy manure estimated by the combination of selective extractions (1), solution (2) and solid-state (3) ^{31}P NMR spectroscopies

P species	Percentage of total P		Identification method
	CD	OD	
Na/K HPO_4^{2-} bonded with H_2O	16.9	24.0	1, 2 and 3
Soluble Ca- and Mg- PO_4^{3-}	24.2	13.5	1, 2 and 3
Fe-related pyrophosphate	5.6	1.7	1, 2 and 3
Moderate soluble hydrogenated Ca-PO_4^{3-}	20.8	-[a]	1, 2 and 3
Moderate soluble tribasic Ca-P minerals	-[a]	35.7	1, 2 and 3
Polyphosphate	2.5	0.9	1 and 2
Soluble DNA	0.8	0.3	1 and 2
Moderate soluble DNA	0.6	0.6	1 and 2
Other diesters (soluble)	2.5	1.7	1 and 2
Moderate Mg and Zn hydrogenated phytate species	6.4	10.0	1, 2 and 3
Other soluble monoesters	5.9	2.8	1 and 2
Moderate soluble other monoesters	3.4	3.8	1 and 2
Al phytate species, some Al phosphate possible	2.0	1.6	1, 2 and 3
Stable divalent metal phytate and phosphate	13.5	6.9	1, 2 and 3
Phosphonates	trace	2.4	3

[a]: Minor moderate hydrogenated Ca-PO_4^{3-} species or tribasic Ca-P minerals might be present.

Data are adapted from He et al. (2009d).

Table 5. Phosphorus functional groups in NaOH-EDTA extracts of soils (receiving 8-10 yr manure applications vs. untreated) as percentage of total peak area of the ^{31}P NMR spectra, or as concentration (mg P kg^{-1}soil; value in parenthesis). The latter was calculated by multiplying NaOH-EDTA extract P by the relevant percentage of peak area

Soil Series	Manure	Extraction Recovery	Inorganic			Organic		
			OrthoP	PyroP	PolyP	Monoesters		Diesters
						phytate	Other	
		(% of total P)						
Bucks	Poultry	84.2	90.4 (1919)	0.54 (12)	1.25 (27)	3.24 (68)	4.60 (98)	0
	Unt [a]	73.7	70.7 (447)	1.09 (7)	0	6.83 (43)	21.40 (135)	0
Chester	Dairy	79.1	88.7 (1958)	0.43 (9)	2.15 (47)	3.80 (84)	4.73 (104)	0.15 (3)
	Unt	84.4	60.7 (443)	1.35 (10)	0	18.34 (134)	18.11 (132)	1.53 (11)
Duffield 1	Dairy	80.4	89.2 (1617)	0.44 (8)	0.48 (9)	4.45 (81)	5.47 (99)	0
	Unt	68.4	59.4 (278)	2.37 (11)	0	11.68 (55)	26.58 (124)	0
Glenelg	Mushroom compost	77.9	45.0 (1707)	4.75 (180)	0	n/a [b]	47.24 (1790)	2.88 (109)
	Unt	68.0	46.1 (189)	4.53 (19)	0	19.42 (80)	29.98 (123)	0
Watson	Swine	69.6	80.9 (640)	1.33 (11)	0	8.41 (66)	9.40 (74)	0
	Unt	97.4	34.3 (213)	4.85 (30)	0	21.95 (136)	36.35 (226)	2.55 (16)
Duffield 2	Poultry	91.2	88.7 (1597)	0.08 (1)	0.71 (13)	4.78 (86)	5.70 (103)	0

Soil Series	Manure	Extraction Recovery	Inorganic			Organic		
			OrthoP	PyroP	PolyP	Monoesters		Diesters
						phytate	Other	
		(% of total P)						
Hagerstown	Poultry	77.9	81.3 (1055)	0.69 (9)	0.71 (9)	7.30 (95)	9.66 (125)	0.36 (5)
Manor	Dairy	72.9	79.4 (467)	1.25 (7)	n/a	n/a	19.40 (114)	n/a
Berks/Weikert 1	Duck	79.9	93.0 (2686)	0.35 (10)	1.51 (44)	1.81 (52)	5.1 (147)	0
Berks/Weikert 2	Poultry	82.7	89.9 (2571)	0.36 (10)	1.24 (35)	4.07 (117)	4.43 (127)	0

[a]: Receiving no manure or fertilizer.

[b]: Not available due to poor resolution.

Data adapted from Dou et al. (2009).

Therefore, soil amendment with moderate levels of poultry manure does not appear to cause accumulation of stable P forms, and that the large HCl-P fraction (31.1% of total P) in poultry manure could act as a source of plant-available P over a short term (e. g. a growing season) (Waldrip-Dail et al., 2009). A pot experiment with rye grass is currently underway to further test the fate of PL P in soils with plant growth.

II.2. Case Study: P Speciation in Heavily Manured Soils

Soils receiving long-term manure applications typically have P built up to levels that far exceed the agronomic optimum required for satisfactory crop production. Such P-enriched soils are often associated with elevated P loss to waters, contributing to accelerated eutrophication of surface waters. Using the Hedley sequential extraction procedure, Sharpley et al. (2004) compared 20 agricultural soils that received poultry, dairy, or swine manure for >10 yr to similar soils receiving no manure or fertilizer. They reported an overall change in P chemistry in the manured soils, with considerable shifts from NaOH extractable to HCl extractable inorganic P in manured soils as compared to the untreated soils. Employing solution ^{31}P NMR spectroscopy, Dou et al. (2009) recently determined P functional groups in some of the samples used in Sharpley's study (2004). The ^{31}P NMR results (Table 5) showed that orthophosphate was the primary form of P in manured (79-93% of total extract P) and untreated soils (33-71%). Pyrophosphate was identified in all soils, whereas polyphosphate was present in seven out of nine manured soils (9-47 mg P kg^{-1} and <2%) but absent in the untreated soils. Organic P accounted for as little as 5.3% of extracted P in one soil that received duck manure, or as high as 60.9 % in another soil that received no P application. Of organic P forms, orthophosphate diesters were detected in only a few soils in relatively small proportions (<3% of extracted P); these soils included both treated and untreated soils. The majority of organic P was in the form of orthophosphate monoesters, with phytate constituting 17 to 43% of the monoesters. Nevertheless, levels of phytate were similar in manured soils (52–116 mg P kg^{-1}, 2–8%) and untreated soils (43–137 mg P kg^{-1}, 6–22%), suggesting little accumulation of NaOH-EDTA extractable phytate despite long-term manure applications. Poultry manure is known for containing high concentrations of phytate. Estimated amount of phytate applied in the relevant soils would be in the magnitude of 300 kg P ha^{-1}

(Dou et al., 2009), translating to about 450 mg phytate P kg^{-1} for the top 5 cm of soils or 150 mg kg^{-1} for the top 15 cm of soils. Clearly, the amounts of phytate detected in these soils (52, 69, 86, 95, and 116 mg P kg^{-1}) were considerably less than the estimated range, suggesting a lack of quantifiable accumulation. The lack of phytate accumulation in manured soils was also reported by He et al (2008b) and Hill and Cade-Menun (2009).

What could have happened to manure-origin phytate in these soils? Part of it may have transferred from the soils in runoff, or it may have translocated downward, and/or leached out of the soil profile. Leaching loss of organic P has been reported (Toor et al., 2003). It is generally believed that phytate molecules have a high affinity for interaction with soil constituents such as (hydr)oxides of Al and Fe, leading to immobilization (Guan et al., 2006; He et al., 2006a; 2006e). However, for P-saturated soils, such immobilization may be limited, rendering manure-borne phytate susceptible to potential leaching or runoff loss. Another possibility is that the phytate was transformed into inorganic orthophosphate or other chemical forms, as suggested in the short laboratory incubation described in section II-1 (Waldrip-Dail et al., 2009). Leytem et al. (2006) also showed that phytate was no longer detectable after a 12-wk incubation of poultry manure-amended calcareous soils while Olsen P in the soils increased. While the observations in both Dou et al. (2009) and Leytem et al. (2006) suggested that degradation of phytate P was occurring in soils, it must be pointed out that profiling of soil P composition based on solution ^{31}P in NaOH-EDTA extracts alone may not always be accurate.

II.3. Case Study: Different Accumulation Patterns of Labile and Stable P Fractions in Pasture Soil after Long-Term Poultry Litter Application

A long-term field experiment in which poultry litter was applied to pasture soils has been maintained for two decades in the Sand Mountain region of north Alabama, USA (He et al., 2008a; 2009b). Five fields received poultry litter (PL) continuously for 0 (control), 5, 10, 15, and 20 years at rates of 2.27, 2.27, 3.63, and 1.36 Mg ha^{-1} yr^{-1}, respectively. Previous work (He et al., 2009b) observed that the P in these five fields had increased with increasing cumulative PL application, suggesting that residual

P from PL had accumulated in these soils. Thus, He et al. (unpublished data) further investigated the P accumulation patterns of this residual PL P by sequential fractionation. In the surface soil (0-20 cm), repeated PL application significantly increased P in all five pools of these pasture soils (Figure 4). However, there were apparently two different impact patterns. The levels of labile P (P_i or P_t) extracted by water or $NaHCO_3$ were more closely related to the number of years of poultry litter application. In contrast, the stable NaOH-extractable P in the surface soil was built up, with a closer relationship to the cumulative applied amount or annual rate. No statistically significant correlation was observed between the HCl-extractable and residual P and the three application parameters. These findings suggest that the PL P accumulated in these soils had undergone transformation and interchanges among the five P fractions or pools. The most apparent change was in the HCl-extractable P_o, which accounted for 21% of total P in PL (Figure 2) but was not present in the HCl fractions of litter-applied soils. The four soluble sequential fractions of soils with 0-yr and 20-yr PL applications were also subjected to solution ^{31}P NMR analysis (He et al., 2008b). The only P species detected in the HCl fractions of both soils was orthophosphate. These observations indicated that PL P accumulated in these soils had gone through transformation and interchanges among the five P fractions or pools in these pastures field as observed in the laboratory aerobic incubation.

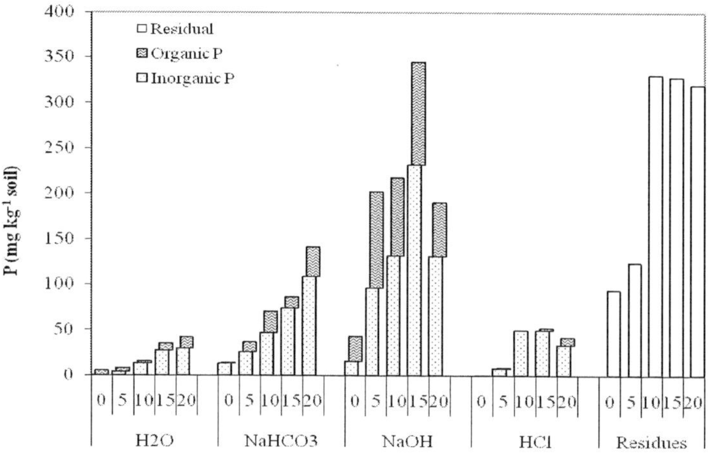

Figure 4. Phosphorus distribution in sequentially extracted H_2O, 0.5 M $NaHCO_3$ (pH8.5), 0.1 M NaOH, 1 MHCl and residual fraction of the surface (0-20 cm) soils collected from the five fields with none (0), 5-, 10-, 15-, and 20-yr of poultry litter application (He et al. unpublished data).

Repeated PL application also increased P in the subsurface horizons (20-40, and 40-60 cm) but at a lower accumulation rate (data not shown). The concentrations of H_2O-extractable P in the 40-60 cm soils after 15- and 20-yr PL application were one-third to one-quarter less than corresponding H_2O-extractable P levels in the surface 0-20 cm soils, indicating there had been little P leaching after long-term PL application in these soils. Although the distribution patterns of P in the two subsurface soils were similar to that of surface soils, the H_2O, $NaHCO_3$ and NaOH fractions of the subsurface soils, especially in the 40-60 cm layer, contained more P_o than their counterpart fractions in the surface soil. Similar to the surface soil, the levels of HCl-extractable P in the two subsurface layers did not correlate to any parameter of the PL applications.

These results not only reveal the transformation mechanism of PL P in pasture soils, but also have implications for environmental impacts. These data clearly indicate that runoff potential (i. e. the level of water extractable P) is more closely related to years of PL application than the annual application rate over the long term. A model incorporating the number of years of PL application could improve the prediction of surface P runoff from long-term PL applied fields (Vadas et al., 2007), thus providing insights to develop soil management and technology for efficiently reducing runoff (or erosion) rate to reverse surface water eutrophication.

REFERENCES

Abbasi, M. K., Hina, M., Khalique, A. and Khan, S. R. (2007). Mineralization of three organic manures used as nitrogen source in a soil incubated under laboratory conditions. Commun. *Soil Sci. Plant Anal*, *38*, 1691-1711.

Ajiboye, B., Akinremi, O. O. and Racz, G. J. (2004). Laboratory characterization of phosphorus in fresh and oven-dried organic amendments. *J. Environ. Qual*, *33*, 1062-1069.

Ajiboye, B., Akinremi, O. O., Hu, Y. and Flaten, D. N. (2007). Phosphorus speciation of sequential extracts of organic amendments using nuclear magnetic resonance and X-ray absorption near-edge structure spectroscopies. *J. Environ. Qual*, *36*, 1563-1576.

Barnett, G. M. (1994a). Phosphorus forms in animal manure. *Bioresour. Technol*, *49*, 139-147.

Barnett, G. M. (1994b). Manure P fractionation. Bioresour. *Technol*, *49*, 149-155.

Beauchemin, S., Hesterberg, D., Chou, J., Beauchemin, M., Simard, R. R. and Sayers, D. E. (2003). Speciation of phosphorus in phosphorus-enriched agricultural soils using X-ray absorption near-edge structure spectroscopy and chemical fractionation. *J. Environ. Qual*, *32*, 1809-1819.

Bowman, R. A. and Moir, J. O. (1993). Basic EDTA as an extractant for soil organic phosphorus. *Soil Sci. Soc. Am. J*, *57*, 1516-1518.

Buehler, S., Oberson, A., Rao, I. M., Friesen, D. K., and Frossard, E. (2002). Sequential phosphorus extraction of a ^{33}P-labeled oxisol under contrasting agricultural systems. *Soil Sci. Soc. Am. J*, *66*, 868-877.

Cade-Menun, B. J. (2005a). Using phosphorus-31 nuclear magnetic resonance spectroscopy to characterize phosphorus in environmental samples., 21-44, *In* B. L. Turner, et al., eds. *Organic Phosphorus in the Environment*. CABI Publishing, Cambridge, MA.

Cade-Menun, B. J. (2005b). Characterizing phosphorus in environmental and agfricultural samples by ^{31}P nuclear magentic resonance spectroscopy. *Talanta*, *66*, 359-371.

Codling, E. E. (2006). Laboratory characterization of extractable phosphorus in poultry litter and poultry litter ash. *Soil Sci.*, *171*, 858-864.

Cross, A. F. and Schlesinger, W. H. (1995). A literature review and evaluation of the Hedley fractionation: applications to the biogeochemical cycle of soil phosphorus in natural ecosystems. *Geoderma*, *64*, 197-214.

Dail, H. W., He, Z., Erich, M. S. and Honeycutt, C. W. (2007). Effect of drying on phosphorus distribution in poultry manure. Commun. *Soil Sci. Plant Anal*, *38*, 1879-1895.

Dao, T. H. and Hoang, K. Q. (2008). Dephosphorylation and quantification of organic phosphorus in poulry litter by purified phytic-acid high affinity *Aspergillus* phosphohydrolases. *Chemosphere*, *72*, 1782-1787.

Dao, T. H., Lugo-Ospina, A., Reeves III, J. B. and Zhang, H. (2006). Wastewater chemsitry and fractionation of bioactive phosphorus in dairy manure. Commun. *Soil Sci. Plant Anal*, *37*, 907-924.

Dick, W. A. and Tabatabai, M. A. (1977). Determination of orthophosphate in aqueous solutions containing labile organic and inorganic phosphorus compounds. *J. Environ. Qual*, *6*, 82-85.

Dou, Z., Toth, J. D., Galligan, D. T., Ramberg, C. F. Jr., and Ferguson, J. D. (2000). Laboratory procedures for characterizing manure phosphorus. *J. Environ. Qual*, *29*, 508-514.

Dou, Z., Ramberg, C. F., Toth, J. D., Wang, Y., Sharpley, A. N., Boyd, S. E., Chen, C. R., Williams, D. and Xu, Z. H. (2009). Phosphorus speciation and sorption-desorption characteristics in heavily manured soils. *Soil Sci. Soc. Am. J*, *73*, 93-101.

Dou, Z., Knowlton, K. F., Kohn, R. A., Wu, Z., Satter, L. D., Zhang, G., Toth, J. D. and Ferguson, J. D. (2002). Phosphorus characteristics of dairy feces affected by diets. *J. Environ. Qual*, *31*, 2058-2065.

Eghball, B., Wienhold, B. J., Woodbury, B. L. and Eigenberg, R. A. (2005). Plant availability of phosphorus in swine slurry and cattle feedlot manure. *Agron. J*, *97*, 542-548.

Griffin, T. S., He, Z. and Honeycutt, C. W. (2005). Manure composition effects net transformations of nitrogen from dairy manures. *Plant Soil*, *273*, 29-38.

Gungor, K., Jurgensen, A. and Karthikeyan, K. G. (2007). Determination of phosphorus speciation in dairy manure using XRD and XANES spectroscopy. *J. Environ. Qual*, *36*, 1856-1863.

Haygarth, P. M. and Sharpley, A. N. (2000). Terminology for phosphorus transfer. *J. Environ. Qual*, *29*, 10-15.

He, Z. and Honeycutt, C. W. (2001). Enzymatic characterization of organic phosphorus in animal manure. *J. Environ. Qual*, *30*, 1685-1692.

He, Z. and Honeycutt, C. W. (2005). A modified molybdate blue method for orthophosphate determination suitable for investigating enzymatic hydrolysis of organic phosphates. Commun. Soil Sci. *Plant Anal*, *36*, 1373-1383.

He, Z., Honeycutt, C. W. and Griffin, T. S. (2003a). Comparative investigation of sequentially extracted P fractions in a sandy loam soil and a swine manure. Commun. *Soil Sci. Plant Anal*, *34*, 1729-1742.

He, Z., Honeycutt, C. W. and Griffin, T. S. (2003b). Enzymatic hydrolysis of organic phosphorus in extracts and resuspensions of swine manure and cattle manure. *Biol. Fertil. Soils*, *38*, 78-83.

He, Z., Griffin, T. S. and Honeycutt, C. W. (2004a). Enzymatic hydrolysis of organic phosphorus in swine manure and soil. *J. Environ. Qual*, *33*, 367-372.

He, Z., Griffin, T. S. and Honeycutt, C. W. (2004b). Phosphorus distribution in dairy manures. *J. Environ. Qual*, *33*, 1528-1534.

He, Z., Griffin, T. S. and Honeycutt, C. W. (2004c). Evaluation of soil phosphorus transformations by sequential fractionation and phosphatase hydrolysis. *Soil Sci.*, *169*, 515-527.

He, Z., Dao, T. H. and Honeycutt, C. W. (2006a). Insoluble Fe-associated inorganic and organic phosphates in animal manure and soil. *Soil Sci.*, *171*, 117-126.

He, Z., Griffin, T. S. and Honeycutt, C. W. (2006b). Soil phosphorus dynamics in response to dairy manure and inorganic fertilizer applications. *Soil Sci.*, *171*, 598-609.

He, Z., Senwo, Z. N., Mankolo, R. N. and. Honeycutt, C. W. (2006c). Phosphorus fractions in poultry litter characterized by sequential fractionation coupled with phosphatase hydrolysis. *J. Food Agri. Environ.*, *4*, 304-312.

He, Z., Toor, G. S., Honeycutt, C. W. and Sims, J. T. (2006d). An enzymatic hydrolysis approach for characterizing labile phosphorus forms in dairy manure under mild assay conditions. *Bioresour. Technol.*, *97*, 1660-1668.

He, Z., Honeycutt, C. W., Zhang, T. and Bertsch, P. M. (2006e). Preparation and FT-IR characterization of metal phytate compounds. *J. Environ. Qual*, *35*, 1319-1328.

He, Z., Tazisong, I. A., Senwo, Z. N. and Zhang, D. (2008a). Soil properties and macro cations status impacted by long-term applied poultry litter. Commun. *Soil Sci. Plant Anal*, *39*, 858-872.

He, Z., Honeycutt, C. W., Zhang, T., Pellechia, P. J. and Caliebe W. A. (2007a). Distinction of metal species of phytate by solid state spectroscopic techniques. *Soil Sci. Soc. Am. J*, *71*, 940-943.

He, Z., Honeycutt, C. W., Cade-Menun, B. J., Senwo, Z. N. and Tazisong, I. A. (2008b). Phosphorus in poultry litter and soil, enzymatic and nuclear magnetic resonance characterization. *Soil Sci. Soc. Am. J*, *72*, 1425-1433.

He, Z., Waldrip, H. W., Erich, M. S. Honeycutt, C. W. and Senwo, Z. N. (2009a). Enzymatic quantification of phytate in animal manure. Commun. *Soil Sci. Plant Anal*, *40*, 566-575.

He, Z., Tazisong, I. A., Senwo, Z. N., Honeycutt, C. W. and Zhang, D. (2009b). Nitrogen and phosphorus accumulation in pasture soil from repeated poultry litter application. Commun. *Soil Sci. Plant Anal*, *40*, 587-599.

He, Z., Honeycutt, C. W., Ohno, T., Wu, F. and Zhang, R. (2009c). Characterization of phosphorus associated with natural organic matter., p. 265-279, *In* F. Wu, and B. Xing, eds. *Natural Organic Matter and its Significance in the Environment*. Science Press, Beijing, China.

He, Z., Fortuna, A., Senwo, Z. N., Tazisong, I. A., Honeycutt, C. W. and Griffin, T. S. (2006f). Hydrochloric fractions in Hedley fractionation may contain inorganic and organic phosphates. *Soil Sci. Soc. Am. J*, *70*, 893-899.

He, Z., Honeycutt, C. W., Xing, B., McDowell, R. W., Pellechia, P. J. and Zhang, T. (2007b). Solid-state Fourier transform infrared and ^{31}P nuclear magnetic resonance spectral features of phosphate compounds. *Soil Sci.*, *172*, 501-515.

He, Z., Cade-Menun, B. J., Toor, G. S., Fortuna, A., Honeycutt, C. W. and Sims, J. T. (2007c). Comparison of phosphorus forms in wet and dried animal manures by solution phosphorus-31 nuclear magnetic resonance spectroscopy and enzymatic hydrolysis. *J. Environ. Qual*, *36*, 1086-1095.

He, Z., Honeycutt, C. W., Griffin, T. S., Cade-Menun, B. J., Pellechia, P. J. and Dou, Z. (2009d). Phosphorus forms in conventional and organic dairy manure identified by solution and solid state P-31 NMR spectroscopy. *J. Environ. Qual*, *38*, 1909-1918.

Hedley, M. J., Stewart, J. W. B. and Chauhan, B. S. (1982a). Changes in inorganic and organic soil phosphorus fractions induced by cultivation practices and by laboratory incubations. *Soil. Sci. Soc. Am. J*, *46*, 970-976.

Hedley, M. J., White, R. E. and Nye, P. H. (1982b). Plant-induced changes in the rhizosphere of rape (*Brassica napus* var.Emerad) seedlings: III. Changes in L value, soil phosphate fractions and phosphatase activity. *New Phytol*, *91*, 45-56.

Hill, J. E. and Cade-Menun, B. J. (2009). Phosphorus-31 nuclear magnetic resonance spectroscopy transect study of poultry opearations on the Delmarva Peninsula. *J. Environ. Qual*, *38*, 130-138.

Hunger, S., Sims, J. T. and Sparks, D. L. (2005). How accurate is the assessment of phosphorus pools in poultry litter by sequential extraction? *J. Environ. Qual*, *34*, 382-389.

Hunger, S., Cho, H., Sims, J. T. and. Sparks, D. L. (2004). Direct speciation of phosphorus in alum-amended poultry litter: solid-state ^{31}P NMR investigation. *Environ. Sci. Technol.*, *38*, 674-681.

Jayasundera, S., Schmidt, W. F., Reeves III, J. B. and Dao, T. H. (2005). Direct ^{31}P NMR spectroscopic measurement of phosphorus forms in dairy manures. *J. Food Agri. Environ.*, *3*, 328-333.

Kleinman, P., Sullivan, D., Wolf, A. M., Brandt, R., Dou, Z., Elliott, H., Kovar, J., Leytem, A., Maguire, R. O., Moore Jr., P. A., Saporito, L., Sharpley, A. N., Shober, A. L., Sims, J. T., Toth, J. D., Toor, G. S. Zhang, H. and Zhang, T. (2007). Selection of a water-extractable phosphorus test for manures and biosolids as an indicator of runoff loss potential. *J. Environ. Qual*, *36*, 1357-1367.

Kleinman, P. J. A., Sharpley, A. N., Wolf, A. M., Beegle, D. B. and Moore, P. A. Jr. (2002). Measuring water-extractable phosphorus in manure as an indicator of phosphorus in runoff. *Soil Sci. Soc. Am. J*, *66*, 2009-2015.

Kruse, J., Leinweber, P., Eckhardt, K., Godlinski, F., Hu, Y. and Zuin, L. (2009). Phosphorus $L_{2,3}$-edge XANES: overview of reference compounds. *J. Synchrotron Rad*, *16*, 247-259.

Leinweber, P., Haumaier, L. and Zech, W. (1997). Sequential extractions and ^{31}P-NMR spectroscopy of phosphorus forms in animal manures, whole soils and particle-size separates from a densely populated livestock area in northwest Germany. *Biol. Fertil. Soils*, *25*, 89-94.

Leytem, A. and Thacker, P. A. (2008). Fecal phosphorus excretion and characterization from swine fed diets containing a variety of cereal grains. *J. Anim. Veterian. Adv*, *7*, 113-120.

Leytem, A. B., Turner, B. L. and Thacker, P. A. (2004). Phosphorus composition of manure from swine fed low-phytate grains: Evidence for hydrolysis in the animal. *J. Environ. Qual*, *33*, 2380-2383.

Leytem, A. B., Smith, D. R., Applegate, T. J. and Thacker, P. A. (2006). The influence of manure phytic acid on phosphorus solubility in calcareous soils. *Soil Sci. Soc. Am. J*, *70*, 1629-1638.

Maguire, R. O., Sims, J. T., Saylor, W. W., Turner, B. L., Angel, R. and Applegate, T. J. (2004). Influence of phytase addition to poultry diets on phosphorus forms and solubility in litters and amended soils. *J. Environ. Qual*, *33*, 2306-2316.

Makarov, M. I. (2005). Phosphorus-containing compounds of soil organic matter: ^{31}P NMR spectroscopic study (a review). *Eurasian Soil Sci.*, *38*, 153-164.

McAuliffe, C. and Peech, M. (1949). Utilization by plants of phosphorus in farm manure: *I. Soil Sci.*, *68*, 179-184.

McDowell, R. W. and Stewart, I. (2005). Phosphorus in fresh and dry dung of grazing dairy cattle, deer, and sheep: Sequential fraction and phosphorus-31 nuclear magnetic resonance analysis. *J. Environ. Qual*, *34*, 598-607.

Murphy, J. and Riley, J. P. (1962). A modified single solution method for the determination of phosphate in natural waters. *Analytica Chimia Acta*, *27*, 31-36.

Pant, H. K., Edwards, A. C. and Vaughan, D. (1994). Extraction, molecular fractionation and enzyme degradation of organically associated phosphorus in soil solutions. *Biol. Fertil. Soils*, *17*, 196-200.

Peak, D., Sims, J. T. and Sparks, D. L. (2002). Solid-state speciation of natural and alum-amended poultry litter using XANES spectroscopy. *Environ. Sci. Technol*, *36*, 4253-4261.

Peperzak, P., Caldwell, A. G., Hunziker, R. R. and Black, C. A. (1959). Phosphorus fractions in manures. *Soil Sci.*, *87*, 293-302.

Preusch, P. L., Adler, P. R., Sikora, L. J. and Tworkoski, T. J. (2002). Nitrogen and phosphorus availability in composted and uncomposted poultry litter. *J. Environ. Qual*, *31*, 2051-2057.

Qian, P. and Schoenau, J. J. (2000). Fractionation of P in soil as influenced by a single addition of liquid swine manure. *Can. J. Soil Sci.*, *80*, 561-566.

Sato, S., Solomon, D., Hyland, C., Ketterings, Q. M. and Lehmann, J. (2005). Phosphorus speciation in manure and manure-amended soils using XANES spectroscopy. *Environ. Sci. Technol.*, *39*, 7485-7491.

Shand, C. A., Coutts, G., Hillier, S., Lumsdon, D. G., Chudek, A. and Eubeler, J. (2005). Phosphorus composition of sheep feces and changes in the field determined by ^{31}P NMR spectroscopy and XRPD. *Environ. Sci. Technol*, *39*, 9205-9210.

Sharpley, A. N. and Moyer, B. (2000). Phosphorus forms in manure and compost and their release during simulated rainfall. *J. Environ. Qual*, *29*, 1462-1469.

Sharpley, A. N., McDowell, R. W. and Kleinman, P. J. A. (2004). Amounts, forms, and solubility of phosphorus in soils receving manure. *Soil Sci. Soc. Am. J*, *68*, 2048-2057.

Shober, A. L., Hesterberg, D. L., Sims, J. T. and Gardner, S. (2006). Characterization of phosphorus species in biosolids and manures using XANES spectroscopy. *J. Environ. Qual*, *35*, 1983-1993.

Toor, G. S., Condron, L. M., Di, H. J., Cameron, K. C. and Cade-Menun, B. J. (2003). Characterization of organic phosphorus in leachate from a grassland soil. *Soil Biol. Biochem*, *35*, 1317-1323.

Toor, G. S., Peak, J. D. and Sims, J. T. (2005). Phosphorus speciation in broiler litter and turkey manure produced from modified diets. *J. Environ. Qual*, *34*, 687-697.

Turner, B. L. (2004). Optimizing phosphorus characterization in animal manures by solution phosphorus-31 nuclear magnetic resonance spectroscopy. *J. Environ. Qual*, *33*, 757-766.

Turner, B. L. and Leytem, A. B. (2004). Phosphorus compounds in sequential extracts of animal manure: chemical speciation and a novel fractionation procedure. *Environ. Sci. Technol.*, *38*, 6101-6108.

Turner, B. L., Mahieu, N. and Condron, L. M. (2003). Phosphorus-31 nuclear magnetic resonance spectral assignments of phosphorus compounds in soil NaOH-EDTA extracts. *Soil Sci. Soc. Am. J*, *67*, 497-510.

Uusitalo, R., Ylivainio, K., Turtola, E. and Kangas, A. (2007). Accumulation and translocation of sparingly soluble manure phosphorus in different types of soils after long-term excessive inputs. *Agric. Food Sci.*, *16*, 317-331.

Vadas, P. A. and Kleinman, P. (2006). Effect of methodology in estimating and interpreting water-extractable phosphorus in animal manures. *J. Environ. Qual*, *35*, 1151-1159.

Vadas, P. A., Gburek, W. J., Sharpley, A. N. Kleinman, P. J. A., Moore Jr., P. A., Cabrera, M. L. and Harmel, R. D. (2007). A model for phosphorus transformation and runoff loss for surface-applied manure. *J. Environ. Qual*, *36*, 324-332.

Waldrip-Dail, H., He, Z., Erich, M. S. and Honeycutt, C. W. (2009). Soil phosphorus dynamics in response to poultry manure amendment. *Soil Sci.*, *174*, 195-201.

Wienhold, B. J. and Miller. P. S. (2004). Phosphorus fractionation in manure from swine fed traditional and low-phytate corn diets. *J. Environ. Qual*, *33*, 389-393.

Wodzinski, R. J. and Ullah, A. H. (1996). Phytase. *Adv. Appl. Microbiol*, *42*, 263-302.

Ylivainio, K., Uusitalo, R. and Turtola, E. (2008). Meat bone meal and fox manure as P sources for ryegrass (*Lolium multiflorun*) grown on a limed soil. Nutr. *Cycl. Agroecosyst*, *81*, 267-278.

Reviewed by Barbara J. Cade-Menun, Agriculture and Agri-Food Canada; Peter A. Vadas, USDA-ARS Dairy Forage Research Center; and Hailin Zhang, Dept. of Plant and Soil Sciences, Oklahoma State University.

R

S

T

U

V

W

X